A Very Speci
Vegetable

By Anna W. Bardaus
Illustrated by Carolina Farías

SCHOLASTIC INC.

Page 31 photos ©: Dreamstime: lotus root (Pichest Boonpanchua), parsnips (Dianazh), radishes (picturepartners); PaulaConnelly/iStockphoto: turnip; Shutterstock: cabbage (monticello), cassava, green beans, and tomatillo (Binh Thanh Bui), eggplant (Max Lashcheuski), pattypan squash (picturepartners), sweet potatoes (wk1003mike), taro (Khumthong).

Written by Anna W. Bardaus. Illustrated by Carolina Farías. Designed by Dorothea Lee.

ISBN-13: 978-1-338-28416-4
ISBN-10: 1-338-28416-9

2 3 4 5 6 7 8 9 10 40 27 26 25 24 23 22 21 20 19

Scholastic Inc., 557 Broadway, New York, NY 10012

Niko was a little boy with a **big** appetite.

What did he like to eat,

you might ask?

He liked to eat...
everything!

But what Niko loved to eat **most**
was a meal with friends.

One day, as he was walking to the market to buy some food,

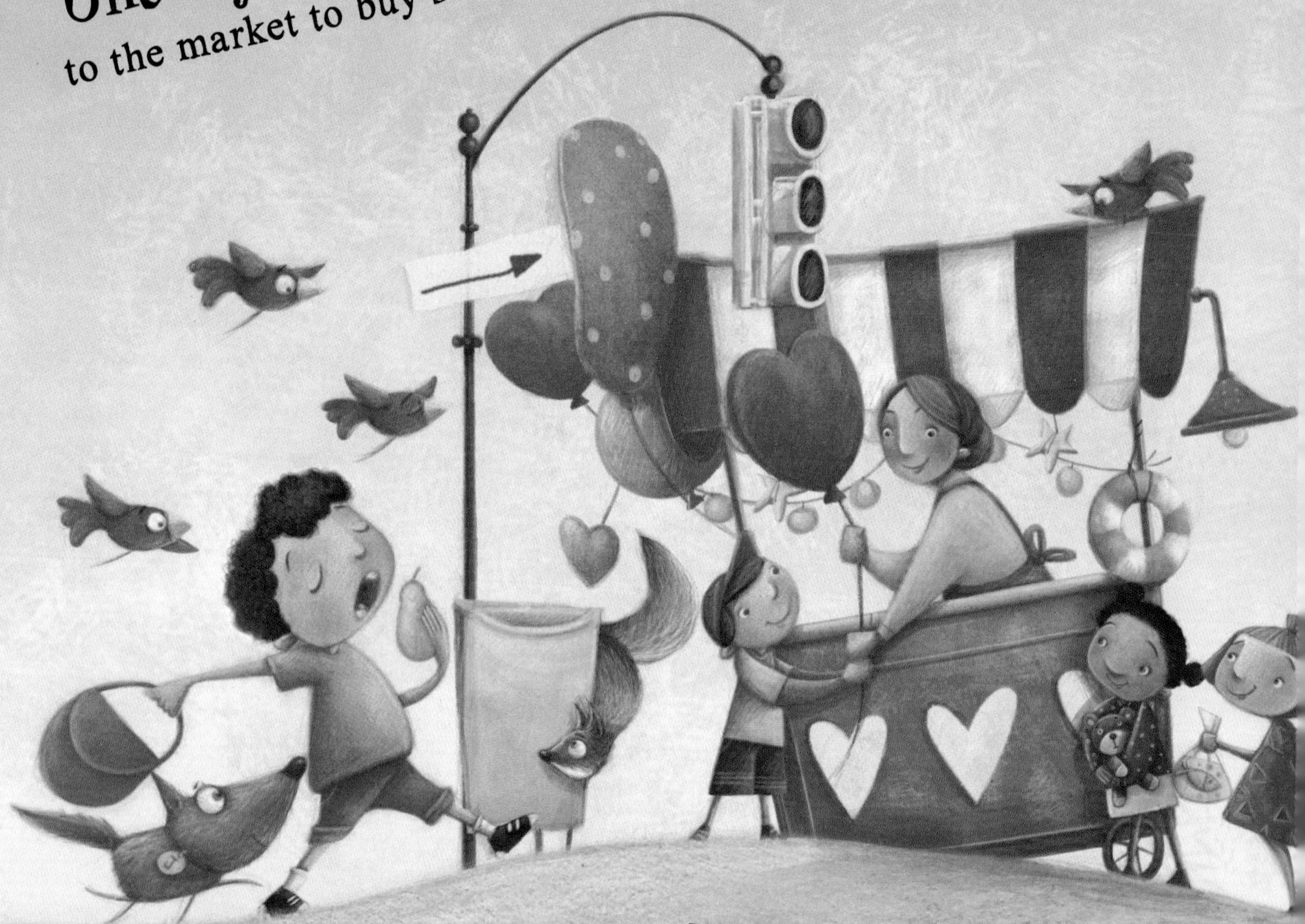

Niko met a traveler with a cart full of **interesting** things.

"I think I know ***just*** what to give you,"
she said to Niko, and handed him a seed.

Carefully, Niko planted his seed in the sunniest, warmest, **happiest** spot he could find.

And as he tucked it into the earth, Niko imagined all of the **delicious** things that might grow...

carrots with a
chompy crunch...

tomatoes full of
dribbly juice...

bright and colorful
sweet potatoes in orange, white, garnet,
and purple...

scalloped pattypan squash
that look like UFOs...

green beans that go
crisp
snap...

long fingers of okra, and the
satisfying pop! their seeds
make between his teeth....

Every day, Niko watered his seed...

and waited...

and waited...

and **waited**.

But as winter turned into spring,
and spring rolled into summer again,
Niko began to believe his seed would **never** sprout.

One warm day,
Niko sat enjoying a **splendidly** messy beet.

And just like that, he noticed something green
poking out of the dirt.

The next day, it had grown...

a lot.

Day 3
Day 5
Day 4
Day 6

By the seventh day,
it was simply **enormous**.

The bigger the plant got, the more people came to see it,
and to guess what might be growing
underneath those **colossal** green leaves.

"I wonder if it's a giant radish," said one. "They have such a hot and tangy **crunch!**"

"Or it could be a **peppery** turnip!" said another. "They both grow underground."

"I wish it were
a tomatillo plant.

The fruits dangle
in such lovely green husks,
like tiny paper lanterns."

"Oh! It could be an eggplant...though...no. They grow from star-shaped flowers."
"It certainly can't be cabbage or asparagus!"
"Kabocha flowers look like golden trumpets!"

"Perhaps a **zesty** parsnip?" asked another.

"A **flaky** potato! Red, or gold, or russet!" said a second.

"I'd **guess** a cassava," a third chimed in.

"I think **surely** it's a taro!" a fourth exclaimed.

"It cannot be crispy, crunchy, creamy lotus root, because they only grow in water.

But, oh my, I do **wish** that it was!"

As the neighbors talked, the plant grew.
Before long, it was as tall as the trees.

"You had better **pull it up now**, Niko," said a man.
"Soon it will be too big to harvest!"

So Niko wrapped his fingers around the great green leaves. They were soft and strong, and made his fingers **tingle**.

But try as he might, the plant would not **budge**.

"We'll help!" offered some children nearby, and they pulled until they had used up their breath.

But the plant stayed **stuck**.

More neighbors joined in, until at last everyone was working together.

And together they **tugged** and they **pulled**...

and they **pulled** and they **tugged**...

And finally, with a satisfying **POP!**,

up came the most **unusual** vegetable the town had **ever** seen.

"Why...that is a **very** special vegetable,"
said someone in the crowd.

And it was a **very special vegetable**, indeed.

Vegetables from Far (and Near!)

Everywhere around the world, vegetables grow. They come in all different shapes and sizes. Here are some vegetables that you saw in this book. How many did you already know? Read on to find out more!

RADISHES come in many shapes, sizes, and colors. The Night of the Radishes is a festival on December 23 in Oaxaca, Mexico. Artists carve sculptures from radishes that can be several feet tall.

TURNIPS were a key food in ancient Rome. Between 77–79 AD, a famous writer, Pliny the Elder, called turnips the most important vegetable of his time.

EGGPLANTS are good luck in Japan. They say that if your first dream of the new year includes an eggplant, you are sure to have good luck for the rest.

GREEN BEANS are popular around the world. People in the city of Blairsville, Georgia, love them so much that they hold a green bean festival every July.

CABBAGE has been grown since around 4000 BC. Over 400 types are grown in over 100 countries today. It's good raw (crunch!) or cooked.

SWEET POTATOES can be red, orange, purple, or white. One of the oldest farmed vegetables, they were first planted in Peru around 8,000 years ago!

CASSAVA, also known as yuca, comes from Brazil, where it is fried, boiled, baked, and powdered to use in "casaba" cakes.

PATTYPAN SQUASH are small with a scallop-shaped edge. They are named after the French word *pâtisson*, a cake that is baked in a scallop-shaped pan.

PARSNIPS get sweeter during cold weather because starch in the root turns into sugar. In medieval Europe, they were used to sweeten desserts.

TOMATILLO was an important ancient Mayan and Aztec crop. Today, it's a key ingredient in salsas and sauces.

TAROS are sacred in Hawaii. In traditional myths, a god named Hāloa-naka was turned into the first taro plant, which was grown to feed the first human being.

LOTUS ROOT is not really a root. It is the stem of the lotus flower. It is a crunchy, starchy food popular across Asia.

Extend the Fun!

Story-Time Tips

1

This story is a mixed retelling of some famous folk tales: **"The Enormous Turnip,"** about a turnip so big it takes a crowd to pull it up; and **"Jack and the Beanstalk,"** about a boy whose magic beans grow a plant all the way to the sky. You can **find more books about these tales at your local library**. Or ask your librarian for some other folk stories to enjoy together.

2

Together with your child, **tell your own version of this story using fruit or another favorite food**. If you know other folktales, tell them your own way, too!

3

Children are more likely to try new foods they help choose. **Make a game of trying new vegetables and fruits**. Once a week or month, let your child choose a vegetable to buy that you've never had before. Take it home, prepare it, and try it together! Be sure to try it more than once. Sometimes children have to see and try a food several times before they like it.

A Very Special ROASTED Vegetable

INGREDIENTS

- 2–3 vegetables from this story (such as beets, sweet potatoes, and carrots)
- salt and seasonings such as garlic powder or rosemary (optional)
- 1 tbsp vegetable oil, such as olive, coconut, or canola

INSTRUCTIONS

1. With the help of an adult, preheat the oven to 400°F.
2. Wash all of the vegetables to remove any dirt.
3. With the help of an adult, cut the vegetables into small cubes.
4. Toss the cut vegetables in a bowl with the oil, salt, and spices.
5. Pour them all onto a baking sheet.
6. Bake at 400°F until slightly crunchy and browned on the edges!